The Tao of Flight Test

Principles to Live By

Mark J. Mondt, II

J.I. Lord
2014

The Tao of Flight Test: Principles to Live By

Copyright © 2014, J.I. Lord

First Printing: 2014

ISBN 978-0692217016

J.I. Lord.
627 S. Clinton St.
Boone, IA 50536 (USA)
www.jilord.com

Contents

Acknowledgements

Popular culture often glamorizes what one person is capable of doing entirely on their own. In real life, however, significant undertakings require the participation of more than one person to bring to a successful conclusion. The production of this book has been no exception. I would like to take this small space to express my gratitude for contributions made in reviewing this work.

Ralph Mohr, a Test Manager with Boeing, has spent many years promoting safety-minded cultures within his flight test teams. Mick Mansfield, a Chief Flight Test Engineer with Bombardier, has researched extensively on the hallmarks of High-Reliability Organizations (HRO). Together their suggestions have led to a more effective delivery of this book's message. Last -- and certainly not least -- my wife Stacy has been the *de facto* copy editor and has helped make the book more readable.

Preface

The Chinese word "Tao" can be generally translated "The Way". In Darwinian fashion, the flight test profession has arrived at a common *way* of doing the job. Early in my career I sensed there was a pattern to this -- a consistent approach -- and eagerly searched for a way to learn all of it as fast I could. Unfortunately for me, every experienced tester I spoke to said this was gained through experience and experience takes time. Supposedly there was no way to systematically transfer this knowledge. Everyone gladly answered my questions but much of the time I just did not know which questions to ask.

> *"Good judgment is the result of experience and experience the result of bad judgment."*
> -- Mark Twain

As time passed I learned The Way. Bits and pieces can be gleaned from older tester's war stories. Elements can be seen in each accident report. But I still remember how it felt to wish for a faster and more systematic way to absorb it all. Certainly new entrants to the profession would be better off if there were better means of indoctrination into The Way.

> *"Learn from the mistakes of others. You can't live long enough to make them all yourself."*
> -- Eleanor Roosevelt

That desire formed the inspiration for this book. My thoughts lie primarily with the new test engineer who finds the profession exciting and wants to learn everything he or she can. Even the experienced tester can use reinforcement from time to time. These few pages represent a small effort on my part to encourage fellow professionals to "do it right" so they can return many times to the ones they love and the ones who love them.

1 —

Look Before You Leap

Some situations are irrecoverable once they occur. The only way of dealing with them is preventing them and the first step in preventing them is to consider the outcomes of choices before those choices are made.

July 27th, 1912
Hempstead, New York

A mechanic was not available to help start the engine so Ernest Stevens, the pilot, started it himself (starting the engine involved manually rotating the propeller while standing on the ground in front of the aircraft). Before Stevens had a chance to get back into the cockpit the aircraft lurched forward, taxiing on its own. All Stevens could do was grab hold of the aircraft's tail. He was seen "flopping ingloriously but valiantly" with his craft until it wrecked into a bank of sand.[1]

Ernest Stevens was not alone. This was all too common a story in his time. While we may consider this a simple problem in our day, it underscores the danger of not thinking ahead. And we are just as likely to fail in forethought as pilots of Stevens' time were. In general, people do not think about the outcomes of an action while they are performing that action. Outcomes generally need to be considered *prior to* the action. Most testers are capable of forethought. The challenge is actually making time to think ahead.

Writing, reviewing or approving formal test procedures seems to be the most opportune way to exercise forethought during a test program. When considering a test procedure, try to envision what it will be like to perform

---

[1] 28 July 1912. "Aeroplane a Runaway: It Has a Wild Race to Destruction, with Stevens Clinging Behind". <u>New York Times</u>.

those actions. Think about what else might be occurring in the airplane or outside it during the prescribed tests. Consider how the test might fail and what those outcomes might be. Do not confine your thoughts to how things "should" work. Keep in mind the people who will be performing your test are human and will need to be balancing other tasks so make sure the test procedure is not asking for too much.

Once the potential negative outcomes of a test are identified, they need to be properly addressed. If the human factors of a test procedure are problematic for a human to consistently perform in the intended fashion then perhaps the design being tested should be modified so the task can be performed consistently. Should this not be possible, consider ways to simulate the task and train for the task before it is performed. When all such mitigations are exhausted -- and the task still appears problematic -- then the test may need to be avoided.

For many test techniques there are data which can provide early warning of irrecoverable problems. When planning tests, consider whether additional data would illuminate the decision to continue or conclude a test sequence. And if such data seems useful then make sure it is available during test conduct. This leads to another potential concern: data overload. One human being can only process and consider so much data at one time. More people may need to be involved in the data monitoring task to ensure that monitoring is performed at a sufficiently-high quality level.

Sometimes the potential outcomes of a test require modifying the test aircraft or the setup for that test in order to be adequately prepared. We wear fire-resistant flight suits to provide some protection should an onboard fire occur. When loss of aircraft control is a concern, helmets are often worn. When runway departure is a concern we make sure people and equipment are placed sufficiently far away from the runway so they do not present another impact hazard. When loss of control in-air is a concern, aircraft attitude recovery systems (e.g., parachutes) can be employed. Skid plates under an aircraft's aft end help mitigate the effects of potential tail strikes during high-attitude takeoff conditions. Many airborne tests carry a minimum altitude restriction to permit an appropriate amount of time for recovery to be effected.

This text cannot possibly hope to detail all the ways in which forethought could and should be exercised in flight test. But the approach is consistent enough across all test programs that it can be stressed here. Consider the potential outcomes of your test. Consider modifications to the design of the aircraft itself to eliminate or control hazards. Consider temporary modifications to the aircraft to mitigate effects of hazards. Consider the utility of specific information in controlling conduct of the test. And, finally, consider changing the manner of the test to avoid risks that cannot be controlled otherwise.

> *"Planning is bringing the future into the present so that you can do something about it now."*
> *-- Alan Lakein*

2 ☰

Fools Rush In Where Wise Men Fear To Tread

> *"Don't be a show-off. Never be too proud to turn back. There are old pilots and bold pilots, but no old, bold pilots."*
> -- E.Hamilton Lee

Forethought and proper planning will reduce but never eliminate risk. As we exercise forethought we are dealing with hazards we can foresee. These are referred to as "known" hazards. But there are also "unknown" hazards -- ones we cannot foresee despite our best efforts. Both categories of hazards can be just as deadly. And when hazards are identified they have their effects or outcomes. Some of these outcomes too can be envisioned ahead of time and some cannot, so even the outcomes can be known or unknown. So how do we deal with the knowledge there is so much we do not know and cannot control?

We cannot elect to do nothing and then still expect to accomplish our mission. We must move forward somehow. This is where it can be helpful to play a test program in reverse. When a test program is concluded, the aircraft's behavior is *generally* known. Prior to the first flight there are many unknowns. Prior to the first ground or laboratory tests even more is unknown. A well-designed test program will move as many hazards and outcomes as possible from the unknown into the known. Once they are known they can be dealt with through redesign, documentation or training.

This is where the general public often misunderstands the true nature of flight test. Popular literature and entertainment programs leave many with the notion that flight testers are somewhat carefree, daring and reckless. Nothing could be further from the truth. What we know is what we have already seen and done. Anything we have not yet done is in unknown territory. And we do not approach unknown territory at a high rate of speed. We approach it methodically. Humility is in order here. Despite all our knowledge we must admit there is much we do not know and we must behave accordingly. History is littered with examples of those who did not give proper respect to the unknown.

"To realize that you do not understand is a virtue;
Not to realize that you do not understand is a defect."
-- Lao Tze

August 15th, 1911
Chicago, Illinois

During the summer of 1910, William Badger witnessed an air meet at Pittsburg. Right then and there he decided to turn his thrill-seeking pursuits from automobile racing to aviation. He practiced for months and eventually obtained his license at Mineola, Long Island from the Aero Club of America. Badger began flying independently (unsponsored) in a red Baldwin biplane. This -- coupled with his fiery red hair -- earned the nickname "Red and his Red".

The August 1911 Chicago air meet was Badger's first public performance. The meet was held at Grant Park and was one of the largest ever meets with an estimated half million spectators involved. Badger attracted a great deal of attention with his recklessness and daring under tricky wind conditions which gave more experienced aviators pause. Prior to the eventual crash, he had soared from within a few feet of the ground to approximately 300 feet above the ground. Then he initiated a steep dive, greatly increasing airspeed then attempted to pull up sharply for another climb while approximately 75 feet from the ground. Experts would later estimate the resulting pressure on the wings at 7,000 pounds, well over that aircraft type's structural limit. Both wings folded and the aircraft dashed to the earth. The wood-and-canvas aircraft structure disintegrated and the engine was partially buried in mud. William Badger was found alive but unconscious, crushed between the radiator and the engine. He was bleeding from both his eyes and his nose and a steel rod from the engine was found to have pierced his skull. He was rushed in an ambulance to St. Luke's hospital but died shortly thereafter. A post-mortem evaluation revealed Badger had also broken his back in several places.[2,3,4]

August 18th, 1911
Aldershot, Hampshire, England

Lt. Theodore Ridge had gained a fair amount of experience piloting dirigibles. He was employed as an assistant superintendent of the Royal Army Balloon Factory at Farnborough. Not long after being checked out in a biplane, he saw a mechanic make several successful flights so he decided he was going to fly that same aircraft himself. The aircraft's designer and an engineer at the balloon factory both told him not to do so but Ridge felt he could manage it. He went for a short flight and upon his return decided to turn the engine off and execute a sharp turn which he had been especially warned to avoid. At this point the aircraft capsized and fell to the ground, pinning the pilot under the debris. Ridge died later that night in the Connaught hospital. An inquest was held with the resulting verdict of "Death by Misadventure".[5,6]

> **"Only a stupid fool is never cautious—
> so be extra careful and stay out of trouble."**
> *-- Proverbs 14:16*

Once you have a healthy regard for what can happen, consider the difference between where you are now in your test program and where you will end up. The gap between the two is your unknown and that gap needs to be bridged carefully, methodically. Sometimes history or simulations and analysis will tell you what dangers might lie in the gap. Sometimes there are total surprises there. Either way the preferred way of

---

[2] 16 August 1911. "Two Airmen Fall To Death. W.R. Badger Crushed by His Engine and St. Croix Johnstone Drowned at Chicago." New York Times, p.1

[3] 19 August 1911. "Two Fatal Accidents At Chicago." Flight, p.727.

[4] Early Aviators: William R. Badger, 1886-1911.
http://www.earlyaviators.com/ebadger.htm

[5] 26 August 1911. "The Fatal Accident at Aldershot." Flight, p.741.

[6] 19 August 1911. "English Aviator Killed." New York Times.

progressing from where you are now (the "known") to where you need to be (the "unknown") is breaking the gap down into smaller steps then taking each step one at a time.

This is generally referred to as the buildup method. Each incremental step should "build" on the previous step and progress you from your beginning to your end. The buildup method should proceed from less perceived risk toward greater perceived risk. Some specific examples are helpful.

One example is envelope expansion, particularly the maximum airspeed portion of the envelope. Go too fast too soon and you might lose the aircraft due to flutter. History has also shown the aircraft could be lost due to a high-speed stability or control issue. As you progress from your starting speed to your endpoint maximum speed you progress in increments primarily in airspeed. The next increment is not attempted unless results at the previous increment suggest the next increment will be okay. Extrapolating across 10 knots or 0.03 Mach can be done with greater accuracy that extrapolating across 200 knots or 0.4 Mach.

Flutter is somewhat unique in that changes in test outcome from acceptable to unacceptable response can occur very quickly. Even then there is a rate of progression, a certain growth rate in bad effects as a function of the chosen airspeed change. With any one increment in airspeed there is only so much response growth that will occur. The greater the increment the more exposure you have. The smaller the increment the less exposure you have. Obviously you do not want to have too big an increment. But making the increment too small could require too many test points and take too long to complete. Although you are venturing into the "unknown", do not do so blindly. Put all the tools in your toolbox to work helping you decide how to move toward that unknown. In the case of flutter those tools would include past experience, dynamic modeling, analysis, and even wind tunnel testing.

Approaching each type of unknown can have its own unique progression method but the general method remains the same. This works numerically where airspeed or load factor might be the increment basis. But it also works categorically for systems testing where nominal (all systems operating) cases are examined before simulated failure cases. Or it could progress from normal usage of a system to worst-case maximum usage of the system. It could also involve starting with more hospitable external factors before subjecting the system to more harsh circumstances (e.g., temperature extremes). The approach is the same whether the increment basis is numerical or categorical. Either way, you divide the known-unknown gap into manageable increments to control the growth rate in potential negative outcomes.

Choosing the buildup method is the most important aspect of progressing toward the unknown. Sometimes, however, the incremental step you have to take is large enough that different control methods need to be used. In the example of stall testing, you can progressively lower your airspeed but sooner or later you need to stall the aircraft, and that is typically an all-or-nothing thing. Even then there are known ways of making stalls worse and making them better. The stall can be approached at varying airspeed decay rates. Greater decay rates can lead to more dramatic results than lower rates. Effects of the stall can be exacerbated if the aircraft is not in trim prior to the approach to stall.

> *"Few things are brought to a successful issue by impetuous desire, but most by calm and prudent forethought."*
> *-- Thucydides*

This is where the tools of the abort or knock-it-off criteria -- along with pre-established recovery tools, methods and procedures -- can be employed. If you cannot totally control an increment then it is helpful to have pre-established limits beyond which you will not allow yourself to go. The manner of flight crew response when reaching those limits should be established ahead of time so the response can be reflexive and immediate. For the stall example there could be an angle-of-attack limit and if that limit is reached before expected stall characteristics are observed then the recovery procedure should be triggered.

The recovery that gets triggered needs to be planned carefully -- just like the planning behind the increment method and knock-it-off / abort criteria. For the stalls this would typically entail pointing the nose down until airspeed increases to a specified amount to ensure normal handling characteristics. Once normal handling is restored, power can be added and the aircraft can climb back to original altitude. Obviously then some altitude is lost in a "normal" stall. And if the stall is abnormal a great deal of altitude could be lost. So this test carries a setup requirement with minimum altitude to ensure the recovery method can actually be accomplished. And, since one of the negative outcomes could be departure from controlled flight from which the airplane is unable to recover, another setup requirement for an attitude-arresting device should be considered. Again, the recovery approach should be established with all the tools in your toolkit. Use past experience to establish which method creates the quickest separation from danger zones. Use simulation to provide some guidance if past experience does not exist.

Like the increment method, your recovery methods and knock-it-off / abort criteria can be used with categorical variables in systems testing too. Perhaps a system configuration change is only allowed to exist for a specified time interval before employing the recovery criteria of going back to the starting configuration (e.g., electrical power system simulated failure conditions). Like you do with the increment method, bring all the tools in the toolkit out when establishing knock-it-off or abort criteria. Use simulation and modeling results. For systems, use bench tests and simulation rig results. Use predictive analysis where it is available. Use these things to establish when or where you expect to see what you are looking for. Then establish a limit for yourself that does not go too far beyond that point.

Few test programs can be considered "exotic" -- pushing our collective frontiers or defining the state of the art. The *general approaches* mentioned here work even for these exotic test programs although the *specifics* will have to be established for the first time. For the rest of us, we have even more tools in the toolkit. Much of the industry uses well-established test methods. The Flight Test Safety Database (http://ftsdb.grc.nasa.gov/) is an excellent resource when examining how past programs have performed certain tests.

If you are part of a test organization that has a long history, then examine past programs' test procedures when considering how to test your program. Even if you are part of a relatively new organization you can avail yourself of literature in the public domain to learn how others in the past have approached the same kind of tests you need to accomplish.

> *"Prepare for the unknown by studying how others in the past have coped with the unforeseeable and the unpredictable."*
> -- *Gen. George S. Patton* (U.S.)

The primary point here is that you do not have to (nor should you) take a blind leap of faith into the unknown. The unknown can and should be approached in a controlled manner. The flight test industry has a "Tao" -- a "way" -- of approaching the unknown and it has served many of us well.

3 三

Brief What You Intend To Fly, Fly What You Brief

No amount of prior planning helps if you do not stick to that plan.
And since no one person has complete knowledge, no one person can
appropriately set the plan -- not even the test pilot! For these reasons,
flight testers from long ago began the tradition of briefing the planned
tests prior to the test. These briefs accomplish several things all at
once.

They clearly define what intentions are for the test. They compare
intended tests with the configuration or maintenance status of the
aircraft. They make use of several different perspectives on the
intended tests and whether plans are sufficiently mature to attempt
those tests. Oftentimes the information or perspectives that surface
during these briefings will cause some changes to be made to the test.
Many times the changes introduced at a brief result in postponement
of tests which simply are not yet ready.

> *"Without good advice everything goes wrong—
> it takes careful planning for things to go right.*
> *-- Proverbs 15:22*

Having the pretest brief and making any necessary changes to the plan
results in an additional layer of safety. This layer's relative thickness
does not matter if you take yourself out from under it later!

Early 1950's
Wright Field, Dayton, Ohio

*Early operation of the P-80 aircraft revealed higher-than-expected
fuel consumption rates so its useful range was less than expected.
Wingtip tanks were fitted but flutter incidents began to occur. The
Air Force decided to study the problem using lead weights in the tip
tanks instead of actual fuel. The test method involved elevator
impulses for exciting symmetric modes and aileron impulses for*

exciting asymmetric modes. Both the wings and tip tanks were instrumented with strain gauges and an accelerometer. Telemetry was apparently not used since recording only occurred onboard the aircraft and analysis was conducted only after each flight. The test plan called for examination of a limited speed range during each flight followed by analysis of the data which would lead to determination of the speed increment allowed on the subsequent flight.

Testing carefully progressed in this manner for several flights. This caused the pilot to think the test point progression was unnecessarily slow. Instead of stopping at the established limit speed on one flight, the pilot decided to obtain a few more speed increments. A very large amplitude wing flutter suddenly developed which violently threw the pilot around the cockpit. Through extreme effort he was able to jettison the tip tanks approximately 3 seconds after the oscillations began. The flutter stopped and the pilot landed the aircraft but the wings were so badly damaged they could not be repaired. Later analysis of the data recorded on the aircraft revealed an approach to flutter that would have been able to be predicted had the pilot complied with the limit speed. While the motion amplitude was small (leading the pilot to conclude nothing untoward was happening), data revealed decreasing damping characteristics.[7]

So the unnamed pilot decided he did not need to abide by the limitations placed upon him by people back on the ground. He nearly paid for it with his own life. The pilot did not appear to totally disregard or have disdain for their expertise. While his thinking was perhaps incomplete, he had what he believed was a good reason to bypass the limitations -- he did not perceive any change in the aircraft's response characteristics.

---

[7] Tolve, L.A. "History of Flight Flutter Testing". Flight Flutter Testing Symposium p 103. Washington, DC, May 15-16, 1958, NASA SP-385, pp.159-166

> *"Consult your friend on all things, especially on those which respect yourself. His counsel may then be useful where your own self-love might impair your judgment."*
> -- Lucius Annaeus Seneca

When it takes so much time and energy to get a test flight started, it can be extremely tempting to improvise when unforeseen circumstances thwart the original plan for that flight. Do not improvise! Do not take yourself out from under the safety layer the pretest brief provides. Wanting to get some value from the flight is natural and can be appropriate when balanced with other factors. If operational flexibility is needed during test, then brief ahead of time what degrees of flexibility are intended. Then if the wider test team finds this agreeable you may take advantage of those provisions later and still be doing so under the briefing's safety layer. Sticking to the plan is the first and most obvious battle.

Properly forming and vetting that plan is a less obvious but no less important battle. No one person can possibly know everything that is needed to be known to conduct flight tests effectively and safely. The viewpoints brought to bear in a briefing should be as full a set as possible. With differing perspectives come differing responses to the briefed plan. When some people find the plan acceptable others may not. No one person should be free to overrule any other person in the brief and everyone in the brief needs to feel free to speak up. For this reason do not squelch the input of others even when that input does not initially seem reasonable to you. Humility is in order here. No one knows everything necessary and that includes you. Do what it takes to bring all possible wisdom to bear on your plan before you embark on that plan. Be open to others' input. Take the time and energy to solicit that input too.

November 20th, 1940
Los Angeles, California

Contract test pilot Vance Breese successfully flew the first 7 flights of the North American NA-73X (serial NX19998), an early variant of what would later become the venerable P-51 Mustang. The eighth flight was flown by North American's Chief Test Pilot, Paul B. Balfour. Breese would later claim to have made a bet with North American executives that Mr. Balfour would crash the prototype on his first flight. The aircraft's designer, Edgar Schmued, offered to show Balfour around the airplane and go through the routine of a takeoff and flight. To this Balfour apparently replied that the NA-73X was like any other airplane and that he would not need a routine checkout.

Ground crew started the prototype's engine at 5:40am and allowed it to warm up to a normal operating temperature. When the engine was restarted just prior to the flight Olaf Anderson, the airplane's mechanic noted it "was a little hard to start". Balfour took off from Mines field at about 7:10am. After making two high speed passes over Mines Field, Balfour apparently forgot to put the fuel valve on "reserve" and ran out of fuel on the third pass. He conducted an emergency landing in a freshly plowed field west of Lincoln Boulevard.

The aircraft overturned after the wheels dug in. Balfour, luckily, was not hurt and was able to crawl out of the upside-down wreck. Later on Robert C. Chilton would be hired as the new Chief Test Pilot and Chilton would continue testing developments of the Mustang throughout World War II.[8,9,10]

No sources available to this author illuminate Balfour's thoughts following the incident flight. However odds are good he regretted not taking Schmued up on his offer. So as you approach your own flights, get all the advice you can on your test plan. Then stick to the plan once it is made.

> **"The end of a matter is better than its beginning, and patience is better than pride."**
> -- *Ecclesiastes 7:8*

---

[8] Wagner, Ray. "Mustang Designer - Edgar Schmued and the P-51."

[9] "This Day In Aviation: 20 November 1941."
   http://www.thisdayinaviation.com/tag/edgar-schmued/

[10] Editors. "Mustang", <u>Airpower</u>. Granada Hills, California, July 1985, Volume 15, Number 4, p. 12.

4 四

Trust, But Verify

> *"Faith and doubt both are needed - not as antagonists, but working side by side to take us around the unknown curve."*
> -- Lillian Smith

No one knows everything in a test program. There is just too much knowledge to be had for any one person to have the time to collect it all. As a result, responsibility gets divided between members of the test team and those members at times need to share summaries of their work. This might include conducting physical work on the aircraft, performing some sort of analysis or even setting up a piece of equipment in a particular manner. While it might be tempting to independently verify each and every piece of information another team member brings, sooner or later you will need to trust what you are being told. Consider the following example of information shared between two crewmembers regarding actions performed.

July 26, 1993
Near Wichita, Kansas

A Canadair Regional Jet (CRJ) 100 (Serial 7001) departed controlled flight while maneuvering and crashed. The two pilots and test engineer aboard were fatally injured and the airplane was destroyed. The loss of control occurred during a low-speed, steady-heading sideslip test maneuver at 12,000 feet. The test plan called for the steady-heading sideslip maneuver to be terminated at 15° of sideslip or at activation of the stall warning stick shaker. However, the captain continued past the stick shaker and reached 21° of sideslip when the departure occurred. The captain requested that the copilot deploy the parachute as the airplane descended through 8,000 feet. Shortly after deployment, the captain asked the copilot if the aircraft recovery parachute was deployed and the copilot responded that it was. Data from the flight data recorder indicated that there was no change in the airplane's acceleration when the parachute was deployed, control was not regained and the airplane's descent continued unabated.

A witness reported seeing the parachute fall free of the airplane as it was deployed. The parachute was found 3 miles from the accident site. There was no evidence of damage to the parachute, risers, shroud lines, or shackle. The shroud lines retained many of the packing folds, indicating that they had never reached full extension. Positions of the parachute system's mechanical elements and controls at the time of deployment could not be determined due to the impact and resulting fire from the crash.

Canadair test procedures called for specific setup provisions with the parachute system prior to any test maneuver that might result in a deep stall or spin. Interviews with other company flight test personnel indicated it was likely that the flightcrew believed the test maneuver did not have the potential to result in a deep stall or spin. Based on the evidence, the Safety Board determined the flightcrew performed the test maneuver without requisite setup steps being taken and that the copilot then deployed the parachute in a manner which allowed the parachute to fall free of the airplane without assisting in recovery from the uncontrolled maneuver.

The Safety Board determined that the probable cause of the Byers, Kansas accident was the captain's failure to adhere to the agreed-upon flight test plan for ending the maneuver at the onset of prestall stick shaker, and the flightcrew's failure to assure that all required switches were properly positioned for anti-spin parachute deployment A factor that contributed to the accident was the inadequate design of the anti-spin parachute system, which allowed deployment of the chute with the hydraulic lock switch in the unlock position.[11]

In this case the captain must have accepted the other pilot's verbal response when executing the anti-spin parachute setup checklist. This illustrates just one way in which trust was exhibited during transfer of information. Other ways might involve analysis being performed to

---

[11] National Transportation Safety Board Recommendation Letter A-94-101, http://www.ntsb.gov/doclib/recletters/1994/A94_101.pdf

determine whether a test is safe enough to do. A common example involves the pilots of an aircraft accepting maintenance staff's characterization of the work performed on the aircraft.

November 22nd, 1944
San Diego, California

Six Consolidated Vultee company crewmembers operated a brand new PB4Y-2 Privateer (BuNo 59554) for its first production test flight. Shortly after takeoff the crew heard a loud crack as a skin panel on the left outboard wing separated. The resulting asymmetric lift condition caused a rolling moment from which the pilots could not recover. The airplane immediately went out of control and crashed near a residential area in Loma Portal, less than two miles from where the aircraft lifted off the runway. The wing panel -- described as having been "twisted like a leaf" -- struck the roof of a residence. Firemen and military rescue crews rushed to the scene but the aircraft was already overtaken by fire and the crew were dead.

Courtesy U.S. National Archives

The wing section was recovered and the cause of the separation was quickly discovered. 98 of the 102 bolts which secured the skin panel to internal structure had never been installed. Two workers who were responsible for installing these missing bolts, and two inspectors who had signed off the work as having been properly completed, were fired.[12,13]

As tempting as it may sometimes be, the tester cannot distrust everything they are told. But then these stories underscore the fact that everything cannot be trusted either. Trust too little and you end up spending too much time essentially doing others' work. Trust too much and you leave yourself more susceptible to others' mistakes (and all humans make mistakes). Somewhere in between the two extremes lies a healthy balance and every tester has to strike that balance at points they are willing to live with.

> *"One must verify or expel his doubts, and convert them into the certainty of Yes or No."*
> -- *Thomas Carlyle*

While there are no hard-and-fast rules for when to trust and when to spend time verifying information there are at least some general principles to consider. First, consider the source. As time goes on you learn who is more careful and complete in their work. While those people are not perfect you can place more confidence in what they tell you than you might place in what others tell you. Second, consider whether the information is coming from only one person or whether the information already involves a degree of verification. Though not a guarantee, maintenance work typically involves verification by quality assurance personnel. Critical analyses performed by engineering staff are typically

12 "This Day In Aviation: 22 November 1944".
   http://www.thisdayinaviation.com/22-november-1944/
13 Veronico, Nicholas A. "Failure at the Factory: A Trio of Wartime Incidents Causes the Loss of Four Aircraft and Six Lives." EAA Warbirds of America, March 2005.

cross-checked by additional, senior engineers. Third, consider the
criticality of getting the information wrong. In these cases give yourself
more time so you can investigate further. Examples of this would include
safety reviews of test procedures and the potential outcomes of
performing those tests. These reviews should be planned to occur far
enough in advance of tests to permit the time required to ask questions
and follow up on resulting actions.

A final consideration is the interpersonal relationship factor. If you find
you cannot trust anyone then you need to remove yourself from the
situation. In this case the well is poisoned and people as people will begin
to insulate and protect themselves from you when they know you distrust
them. This almost always means they will not be sharing information with
you freely and that is the last thing you want to have happen. You want
the opposite. You want people to volunteer concerns even if they seem
minor to you because sometime somewhere that concern might not be so
little. And you will want people to be resolute when they stand up and say
their work has been completed. Once you feel you can trust people, let
them see you trust them. Trust is indispensable to a healthy, functioning
test team. Trust cannot and should not be given implicitly. But trust must
typically be given before it can be received.

> *"He who does not trust enough,*
> *Will not be trusted."*
> -- Lao Tze

5 五

Accept No Unnecessary Risk

Flight Test is an inherently risky business. Great energies are summoned in an attempt to minimize the risks -- and rightly so. But no amount of energy will eliminate risk altogether. The only way to eliminate risk is to simply do nothing. Since flight test stands between the expense of development and the potential return on that investment in sales, organizations cannot afford for flight test to stand still. Only when a test article has been deemed completely unsafe to operate should a test program stop completely. So flight test must face risk head-on. The question is not "if" or "whether" but "how".

> *"Take calculated risks. That is quite different from being rash."*
> *-- Gen. George S. Patton*

The key here is making sure sufficient reason for doing the test exists. Engineers always want more data, but if the data is not required then the test is not required. Unfortunately what is "required" to one person may not be "required" to another, so a common basis for establishing requirements is in order. Organizations are required to perform flight tests to ensure the customer is not being presented with an unsafe product and that the customers actually get what they had intended to receive when they committed to purchasing the aircraft. Military and civilian authorities have established complete sets of requirements that potential aircraft must meet before being approved for purchase. Those regulations form the bare minimum requirement set for any test program. Any specific contractual guarantees which require verification should also be included in the requirement set.

Once these requirements are defined, a list of tests can be devised to meet them. For both safety and economic reasons, those tests should be the minimum possible set of tests to satisfy the requirements. Some tests do not directly satisfy requirements but instead feed data to an analysis which is required. And, finally, some tests become required as part of troubleshooting efforts to understand test failures so the design can be corrected to bring about a passing result in the future. If there are any other reasons to perform a test, then the test does not need to be performed. And if it doesn't *need* to be performed then it should not be. Consider the XB-70 story.

June 8th, 1966
Mojave Desert (near Barstow), California

A public relations firm for General Electric (GE) had been bugging the test force director for photos of the XB-70 to include in a brochure for GE's upcoming shareholders' meeting. They wanted a shot with differing General Electric engine-equipped aircraft in formation with the XB-70. The director eventually agreed to the request and obtained approvals from his immediate superiors. A plan was then developed to fly the XB-70's second prototype with a F-4B, T-38A, YF-5A, B-58 and F-104N. A Learjet flown by Clay Lacy was slated as the photo chase vehicle. All that remained was a test mission (on "non-interference" basis of course) at slower speeds the photo chase could keep up with.

The arrival of a new pilot to the test team provided that opportunity. The plan called for 12 subsonic (plus one supersonic) airspeed calibration runs then rendezvous with the other aircraft over the ocean just off the Santa Barbara coastline. The B-58 broke down that morning leaving only 4 aircraft for the formation shot. The rendezvous occurred at 0827 hours with the aircraft arranged in a V-formation. The photo session lasted from 0845 to 0925 hours. This was lengthy for a photo shoot but conditions were cloudy and less favorable for photographs at the original location so the flight repositioned over land where the flight found a slight overcast but clear conditions with very little turbulence. Another F-104, a two-seater with another photographer, received permission to join the formation for a few pictures before returning to Edwards. This second F-104's pilot would later report that the formation looked good, although the two aircraft on the left were not flying as close as the F-104N and the

YF-5A on the right were. The F-104N was estimated to be flying 70 feet to the right and 10 feet below the XB-70's folded-down right wingtip.

The USAF summary incident report would later state that, from this position, the F-104N pilot (Joe Walker) would not have been able to see the XB-70's wing except by looking uncomfortably over his left shoulder. Some sources indicate Walker used the engine inlet ramp and right wing leading edge for visual reference. Regardless of the focal point used, Walker's ability to judge relative motion between the two aircraft was certainly compromised. Test control reported a B-58 (not the intended formation participant) approaching high above in the supersonic corridor which was not deemed a hazard. Several pilots in the formation responded they could see the B-58's contrail but Walker never made such a call. He did, however, comment that he was getting turbulence from the XB-70's wing and that he was trying to hold his position whereupon he repeated his position from the preflight brief stating -- for the record -- he opposed the mission as being too dangerous and having no scientific value.

Suddenly, at 0926, the F-104N collided with the Valkyrie's right wingtip. The Air Force's summary report would later blame vortices shedding from the XB-70's right wing impinging on the F-104's tail reducing its longitudinal trim stability. The F-104 then pitched up violently and rolled inverted, passed over the XB-70 and sheared off most of the larger jet's tail fins which in turned cleaved Walker's cockpit and flight helmet. The F-104N burst into flames and disintegrated almost immediately.

"Midair, midair, midair", came the call from the Test Force Director (Col. Joe Cotton) who was riding backseat in the T-38, "You got the verticals came off, left and right. We're staying with you." The XB-70 maintained straight and level flight for 16 seconds almost as if nothing had happened. But then it rolled slowly to the right and entered an inverted spiral shedding parts and a steady stream of fuel from its broken right wing on its way down. Cotton continued on the radio "The B-70 is turning over on its side and starting to spin. I see no chute. Bail Out, Bail Out!!" One of the XB-70's pilots would successfully eject in the specially-designed escape capsule but the other

pilot did not initiate the eject sequence soon enough and excessive g-forces kept the ejection mechanism from working.

The test program would continue using the slower-speed first prototype for a few more flights but a $4 billion program would eventually be cancelled. As large as the aircraft was, it was said to be worth 10 times its weight in gold so expensive was its design, manufacture and operation. Two highly-skilled and experienced pilots were lost and all of this for a photo in a corporate brochure. With the public fallout from the crash, General Electric eventually chose not to use the images gained from the photo shoot.[14,15,16]

Note the "non interference" agreement to the formation flight. This is a common way for people to deceive themselves into thinking they are not adding risk to a test program. The only way an activity truly does not interfere is if it takes no extra time or effort. Clearly extra time and effort were required to accomplish the formation photo portion of the XB-70 flight.

Even seemingly-required tests need to be evaluated for the potential hazards involved. Sometimes the tests need to be performed differently than originally conceived in order to bring the risk down to acceptable levels (if the risk associated with doing a test a certain way is unacceptable then that risk is still unacceptable even if the test is deemed to be "required.") The requirements themselves can be altered somewhat when the risks involved are judged to be unacceptable. Every program will have some appointed regulatory authority overseeing their work who must approve their test plans. Those regulatory authorities have their own ways of approving

---

[14] Summary Report: XB-70 Accident Investigation. USAF, 1966.

[15] "Secret Heroes: The Crash of XB-70 002" http://area51specialprojects.com/xb70_crash.html

[16] "Midair! Midair!" http://www.thexhunters.com/xpeditions/xb-70a_accident.html

exceptions. These exceptions processes should not and can not be over-used but they do exist and should be used when needed.

Managing risks is a critical function with any test program. Judgment is required and quality of judgment can be uneven from one person or program to the next. Every test team - regardless of technical challenges, size, or experience level - needs to have a systematic and complete approach to take the guesswork out of this all-important task. Regulatory agencies worldwide have recognized this need, and they appear to be converging on processes they require from organizations performing flight test activities[17].[18,19,20]

Those processes will not be detailed here. Ample literature already exists detailing how these processes can and should be carried out. They do, however, have some common elements. The first is identifying the hazards involved with performing each test. Second is identifying the level of risk those hazards entail. Third is assigning mitigations to either minimize the probability or severity of the hazard's occurrence. This process is not always a linear, one-time-through process. If the risk level is too great, then something needs

---

[17] Aircraft Certification Service Flight Test Risk Management Program. U.S. Federal Aviation Administration Order 4040.26B effective January 31st, 2012.

[18] Safety Management Manual, third edition. International Congress of Aviation Organizations (ICAO) Document 9859-AN/474, 2013. http://www.skybrary.aero/bookshelf/books/644.pdf Safety Management Manual, third edition. International Congress of Aviation Organizations (ICAO) Document 9859-AN/474, 2013. http://www.skybrary.aero/bookshelf/books/644.pdf

[19] Safety Risk Management Policy. U.S. Federal Aviation Administration Order 8040.4A effective April 30th, 2012. http://www.faa.gov/documentLibrary/media/Order/8040.4A%20.pdf

[20] Department of Defense Standard Practice: System Safety. MIL-STD-882E effective May 11th, 2012. http://www.system-safety.org/Documents/MIL-STD-882E.pdf

revision -- either the requirement for the test, the method of test
employed, or the setup for the test. If increasing knowledge of the
aircraft would change a previous risk assessment, then that risk
assessment should be revised.

The danger has not passed once you have identified activities which
should or should not be performed. You still need to stick to the plan.
There will always be surprises. There will always be last-minute
requests. But once a process has been established to identify required
and non-required tests then adhere to the process even if it seems to
slow things down. And once you decide that something should not be
done then don't do it. Do not allow external pressure to override your
better judgment.

September 14th, 1912
Chicago, Illinois

*The Aero Club of Illinois organized several events as part of an
aviation meet hosted at Cicero Aviation Field in Chicago. The last
event planned on the incident day was a comparatively long, 20-
kilometer race. With schedule slips during the day, the final event
occurred at twilight which led one of the competitors to make protest
with race officials. George Mestach complained to the officials that it
was too dark to fly safely. He pointed out the race had been placed
at the end of the day's program and that he did not want the race to
start on the verge of night. Frenchman Mestach declared that he
could not see at all. The race officials called the start anyway and
Mestach decided to go through with the race and took off.*

*As Mestach was rounding the pylon at the field's northeast corner in
his Borel monoplane he noted another flyer, Howard Gill (flying a
Wright biplane) closing in on him at a reported 60mile per hour
closure rate. Gill apparently attempted to duck under Mestach and
Mestach pitched up sharply to avoid Gill. Mestach would later
complain of his inability to get a good view of the biplane. The two
airplanes struck in mid-air at a height of approximately 200 feet,
their wings became entangled and both craft plummeted to the ground.*

Both aviators were apparently not wearing seat belts and fell free from their seats.

The 5,000 people who were watching the races rushed to the scene. Race officials and the police tried in vain to keep them back. Ambulances removed the aviators to nearby St. Anthony De Padua Hospital. Gill died en route but Mestach revived a half hour after arriving at the hospital and recounted his version of events to waiting reporters.

"I was forced into making the flight by the insistent demands of the spectators," said Mestach. "It was too dark for safe flying with another machine in the air. When I got on the far turn of the course I heard the whirring of Gill's machine, and saw him bearing down on me. I tried my best to avoid the collision, but could not get my machine to answer the levers quick enough. Then came the wreck. That is all that I remember." [21,22]

> **"Be wary of the man who urges an action in which he himself incurs no risk."**
> -- Lucius Annaeus Seneca

September 1ˢᵗ, 1911
Norton, Kansas

For several days J. J. Frisbie had been giving exhibitions at the Norton County Fair in his Curtiss biplane. The previous day his machine had malfunctioned and his craft "fell" from 40 feet. Luckily Frisbie escaped with nothing more than bruises but he was skeptical about his machine's ability to withstand further flying so he announced that he would not be going up again. When this statement was communicated to the crowd, the spectators hooted and shouted "faker" and refused to listen to explanations. Frisbie then announced that he would go ahead and attempt the flight since he did

---

[21] 15 September 1912. " Aviator Killed in Midair Collision: Howard Gill Hurled 200 Feet to Ground in Crash at Chicago Meet." New York Times

[22] 21 September 1912. "Two American Fatalities." Flight, p.861.

not want the crowd to be left with the impression he was not willing to do his best.

With his wife, son and little girl in attendance, Frisbie took off and climbed to 100 feet. The flight was proceeding without event until he attempted a turn and lost control. The spectators could see he was trying hard to control the machine to no avail. As the aircraft approached the ground, one of its wings glanced off a barn then struck the ground pinning Frisbie underneath its engine which crushed his left side and chest. He is said to have lived for about an hour following impact. Frisbie's wife was overcome with grief when she saw her husband fall. She bitterly denounced the spectators for having forced Frisbie to make the flight with a disabled machine.[23,24]

Trust your instincts. If you have a question whether something should be done -- just don't do it. Back up and re-evaluate. Do not attempt the test again until your judgment is altered by additional knowledge or the aircraft configuration is improved or the procedure is modified. At every point in time you must believe the risks being incurred are at an appropriate level.

---

[23] 02 September 1911. "Crowd Goads Airman to Flight and Death: J. J. Frisbie Goes Up in Crippled Machine Because Kansas Spectators Call Him a Faker." New York Times.

[24] 09 September 1911. "A Disastrous Weekend." Flight, p.790.

6 六

Follow the Truth Wherever it Leads

A good tester is an un-biased reporter. The schedule and cost implications of labeling a test result acceptable or unacceptable must *not* be considered when making the determination. Of course this is easier said than done. Failing tests require repeats at the very least and often entail redesign which can lead to significant growth to both cost and schedule. When the success or failure of the entire program appears to hinge upon the test outcome, even the most experienced testers can be sucked into the desire to make the test "pass".

April 2nd, 2011
Roswell, New Mexico

In preparing for the G650 field performance flight tests, Gulfstream considered ground effect when making initial predictions of the airplane's takeoff performance capability. In the process, errant assumptions were made resulting in estimated stall angle of attack (AOA) values being set higher than actual in-ground-effect stall AOA. As a result, the airplane's stall warning activation levels were set too high, and the flight crew received no tactile or visual warning before the actual stall occurred. Along with the mistakenly-estimated higher stall AOA, predicted V2 speeds were also lower than actual speeds. Actual takeoff distance, consequently, were longer than predicted takeoff distances. The predicted-value errors were not made known within the test team until after the accident flight. The test team did, however, experience results which could have led them to question the predictions.

During an internal meeting held to discuss issues from early field performance testing at Roswell, Gulfstream executives were informed that the recorded $V2$ speeds were high. If these results were to hold during final testing phases, the field length needed for takeoff would be longer than the program's takeoff performance guarantee (6,000 feet $\pm$ 8 percent) to customers and Gulfstream would be subject to performance penalties from customers and would not be permitted to operate from some smaller runways. At that time the test team believed changes to the takeoff technique could improve the results. During subsequent flights, the team experimented with different takeoff rotation techniques and rotation speeds to try to eliminate the $V2$ overshoots. Takeoff technique changes included: (1) adding 2 knots to the V_R speed schedule for a given thrust-to-weight ratio while keeping V_{LOF} and V_2 the same and (2) increasing the pitch angle beyond the target pitch angle as soon as the airplane lifted off (instead of holding the target pitch angle until above 35 feet). In addition, changes were made to the control column input used to initiate rotation. As the testing progressed, the abruptness and magnitude of the rotation-initiating control column input was increased. At one point the FTE asked the pilot whether he could convince FAA certification officials that the rotation technique being explored was a "normal technique." The pilot responded that the technique would have to be modified "slightly."

The test team found that an abrupt column pull force of about 70 to 75 pounds was the most successful in reducing the magnitude of the V_2 overshoot. (maximum permissible column force per FAA regulations was 75 pounds) The test team also found if the flying pilot rotated rapidly (peak pitch rates between 6.1° and 8.5° per second) to the 9° target pitch attitude and then exceeded 9° shortly afterward, V_2 overshoots (and V_2 + 10 knot overshoots for all-engines-operating takeoffs) could be reduced to within 3 knots of the target speeds. According to Gulfstream's released test plan, the required tolerance for the target $V2$ speed was $\pm$ 2 knots.

During the brief to the accident flight 153, the FTE stated the test would be discontinued if the rotation angle of attack exceeded 11°. Recovery procedures were to include lowering the nose and increasing power. Testing initially comprised flaps 20 AEO and one-engine inoperative (OEI) continued takeoff points and would then progress to flaps 10 points. After one of the test points was completed, the pilot expressed concern about capturing V_2 + 10 knots at 35 feet, stating, "the only thing I can say is you're not gonna be at 9 degrees very long if you want to catch V_2". The FTE concurred. After a few more test runs, the pilot noted that he would have to aim for 15° or 16° of pitch to capture V_2. The pilot also indicated that he was doing "a nice smooth ramp" and "I'm not doing that jerk stuff...it just doesn't work. That's not the way they're going to fly the airplane, and I don't think the FAA's gonna like it either...it's such a great flying airplane, you shouldn't have to abuse it to get [it] flying." The FTE and pilot concurred on the need for a near-continuous pull, not pausing very long at 9 degrees pitch. Following a subsequent test run, the FTE stated "I think that's it," and the pilot stated, "we're done, I think we caught it there...we must be onto something now."

The next test points were attempted at flaps 10. The FTE indicated the column force was good and the pitch target had been "nailed." After the test run, however, the FTE indicated the V_2+10 target had been exceeded by 12 knots. The pilot stated, "there's very little time at 9 [degrees]...you wanna try one more and I'll just pause at 9 [degrees] and just keep going?" The FTE agreed with the pilot's plan and the pilot continued "I'll capture it and boom we're back into it...it's almost a continual rotation. You can target 9 [degrees], but you don't want to hang out there very long...now we're into kind of a technique thing here in how we're gonna do this. The FTE replied "that's what I was hoping, [to] just spend today just to get something we like." During the next run the pilot and FTE agreed the targeted pitch was good and the speed target was only exceeded by 7 knots. Another FTE in the telemetry trailer then indicated pitch had reached the knock-it-off of 11° a half-second second after liftoff. The pilot responded "we didn't pause very long at 9 [degrees]. We're *trying to capture that V_2 at 35 [feet], so…it's just not there very long, so I think that's what you were seeing."*

On the next run the V_2 target was exceeded by 9 knots. After the run, the FTE commented about the delay in liftoff that occurred after achieving the target pitch angle. The pilot flying replied, "well we're pausing, because we're tryin' to do this capture, and I think we're getting too focused on that 'cause if you have a real engine failure, the guys aren't gonna be lookin' at nine degrees, they're gonna be lookin' at tryin' to get to V_2." For all of the day's runs, V_2 target exceedances ranged from 4 to 7 knots for flaps 20 points to between 7 and 12 knots for flaps 10 points. The last, accident run proceeded as follows:

Time	Event
0933:17 to 0933:35	*The airplane lined up with runway heading on center-line, power was set for takeoff, and the brakes were released. During acceleration, wheel was held at about 1° to 2° left, and the airplane's roll angle remained within 1° of level. The pilot flying maintained a rudder pedal deflection of about 0.5 inch to the right.*
0933:36 to 0933:38	*The right thrust lever was pulled back at the engine failure speed briefed for the takeoff (105 knots), and the thrust from the right engine decreased. The pilot flying moved the rudder pedal to about 1.4 inches left and modulated the input to maintain the runway heading.*
0933:45 to 0933:47	*As the airplane was accelerating through 123 knots, the pilot monitoring called "standby, rotate." As the airplane accelerated from 125 to 127 knots, the pilot flying pulled the control column 6° aft. A 0.5°-per-second right yaw rate developed.*
0933:47 to 0933:50	*Pitch rate reached a peak of 6° per second and then relaxed to about 1° per second as the pitch angle was increasing through about 9°. A right roll rate started to develop. The pilot flying's control wheel input increased from about 1.8° left to about 11.9° left. The left MLG tires lifted off the runway at 0933:48.8.*

Time	Event
0933:50 to 0933:52	The right MLG tires lifted off the runway at 0933.50.3. The stall on the right wing occurred at an AOA of 11.2° at 0933:50.5. The pilot flying moved the control wheel from 11.9° left to 22.6° left. The roll rate to the <u>right</u> increased to about 4.9° per second. The yaw rate to the right started increasing continuously, passing 2° per second at 0933:52. The pilot flying increased the left pedal deflection.
0933:52.0 to 0933:52.7	The AOA reached 12.4°, and the stick shaker activated. Both pilots exclaimed "whoa, whoa". The pilot flying pushed the control column forward abruptly, moving it from 2.5° aft to 1.2° forward. The roll rate increased to a peak of 9.6° per second to the right. The pilot flying moved the control wheel abruptly from 26.5° left to 60° left (full deflection). The roll angle reached 15.5° right wing down and was increasing. The pilot flying moved the rudder pedal full left.
0933:52.5 to 0933:53.3	The pitch angle decreased from 12.9° to 11.5°, and the AOA decreased from about 12.7° to about 11.5°. The airplane's right wingtip contacted the runway at a roll angle of 13.4°. The roll rate then reversed rapidly from 9.6° per second to the right to 1.3° per second to the left. Increased background noise was noted on the voice recorder. The pilot flying pulled back on the control column with about 38 pounds of force, moving the column to about 4° aft. The stick shaker de-activated.
0933:53.5 to 0933:53.8	The right thrust lever was advanced to match the left thrust lever. Right engine power started to increase. The stick shaker fired and the TAWS bank angle alert annunciated
0933:54 to 0934:00	The pilot called "power, power, power" and relaxed the control column to about 1.5° aft and then pulled back again with more than 60 pounds of force, moving the column to about 7.5° aft. The column remained aft, with the pilot flying pulling between 60 and 110 pounds of force. The yaw rate to the right increased to 9.5° per second.

Time	Event
09:33:57.4 *to* *09:34:00*	*The copilot exclaimed "no, no, no, no". The roll angle increased to 32° right wing down. The bank angle alert sounded again. The roll rate then reversed, and the roll angle decreased to 17° right wing down. The pilot flying's last words were "ah, sorry guys."*

Courtesy U.S. National Transportation Safety Board

<u>*Probable Cause*</u>

"The National Transportation Safety Board determines that the cause of this accident was an aerodynamic stall and subsequent uncommanded roll during a one engine-inoperative takeoff flight test, which were the result of (1) Gulfstream's failure to properly develop and validate takeoff speeds for the flight tests and recognize and correct the takeoff safety speed (V2) error during previous G650 flight tests, (2) the G650 flight test team's persistent and increasingly aggressive attempts to achieve V2 speeds that were erroneously low, and (3) Gulfstream's inadequate investigation of previous G650 uncommanded roll events, which indicated that the company's estimated stall angle of attack while the airplane was in ground effect was too high. Contributing to the accident was Gulfstream's failure to effectively manage the G650 flight test program by pursuing an aggressive program schedule without ensuring that the roles and responsibilities of team members had been appropriately defined and *implemented, engineering processes had received sufficient technical planning and oversight, potential hazards had been fully identified, and appropriate risk controls had been implemented and were functioning as intended."[25]*

Hindsight is 20/20 as the saying goes. Unfortunately foresight is not so clear. Exercise humility and avoid the temptation to consider yourself above others' mistakes when reading their accident reports. Odds are those professionals knew the importance of remaining unbiased and believed they were. The only defense the rest of us have against this is to learn from others' unfortunate mistakes and recognize when our situation has similar characteristics. When you find yourself in similar situations assume you have become biased and invite external perspective to assess whether you have become emotionally invested in the test outcome.

> *"People can foresee the future only when it coincides with their own wishes, and the most grossly obvious facts can be ignored when they are unwelcome."*
> *– George Orwell*

Polluting one's perception of test outcomes with one's hopes can result in two unwanted outcomes. Like the G650 crew it can alter our perceptions of reality and cause us to enter into unnecessary risk. Perhaps more insidious is the urge to downgrade the impact assessment from unanticipated test results, thereby passing the problem on to the end user following flight test. In either case the good tester needs to avoid presaging results. We are all human and everyone likes to report good news. While making people happy in the short term is fun, serving end-users in the longer term is the reason we exist. Our own longevity is a legitimate consideration too. Check hopes for outcomes at the door, and

---

[25] "Aircraft Accident Report: Crash During Experimental Test Flight Gulfstream Aerospace Corporation GVI (G650), N652GD Roswell, New Mexico April 2, 2011." National Transportation Safety Board (NTSB)/AAR-12/02 Adopted October 10, 2012. http://www.ntsb.gov/doclib/reports/2012/AAR1202.pdf

be prepared both for success and failure of each test. And once a test is complete, ensure test reports contain a clear and unbiased accounting of the results.

> *"Science is a way of trying not to fool yourself.*
> *The first principle is that you must not fool*
> *yourself, and you are the easiest person to fool."*
> — Richard Feynman

7 七

Listen to What the Airplane is Trying to Say

Test programs have a mind-boggling number of details to manage. All kinds of schemes are employed to ensure no detail is missed when planning and conducting individual test flights. This detailed focus can make it hard to "connect the dots" or see the proverbial forest instead of the trees. However, trends between tests and relationship of individual tests to the rest of a program (before and after that test) *must* be kept in view. This is easier said than done.

April 26th, 2003
near Loma Alta, Texas

During the flutter test program, limitations were placed on the degree of aileron trim usage due to deficiencies with the trim motor nearer the movement limits. Complicating the issue was the need for greater amounts of left-wing-down trim with increasing airspeeds. This led to a temporary speed limitation of 250 knots. Flow visualization studies were performed which revealed large regions of shock-induced separation above 0.81M. An internal safety review board found that full left-wing-down trim (along with hands continually holding the wheel) would be needed to complete tests above 320 knots (calibrated).

Flight 230 was the first flight of the higher-speed flutter test program and called for test points between 0.844M and 0.864M. During the preflight brief, the test team decided to continue with the planned flutter test only if the amount of wheel force required to maintain wings level at full left trim were "small". The flutter test consultant Designated Engineering Representative (DER) indicated his analysis would be able to be completed if aileron control pulses were sufficiently greater than the steady-state wheel force. At one point during the flight all of the left trim was required and rudder was even used to assist. On the following test point the aircraft experienced an uncommanded left-wing-down roll during which the pilot noted a "rumble". When the pilot noted differences between his indicated airspeeds and the speeds being reported by the chase aircraft the decision was made to terminate the test. Following the flight, the pilot realized that he had incorrectly set up the airspeed display in the test airplane, and was actually flying faster than his airspeed indicated.

The briefing for flight 231 was conducted by telephone between the crew at the airfield in San Antonio and the telemetry van personnel. Test points were briefed between 0.884M and 0.894M. The amount of airspeed increase over the points of flight 230 were apparently due to flight 230's speeds having been higher than planned following the air data setup error. The flutter consultant repeated his statement from prior discussions that flutter testing could still be completed with full aileron/elevator trim even if the pilot had to hold aileron/elevator force to steady the airplane. He indicated the final decision on acceptability of the control forces rested with the pilot. Flight 231 proceeded normally from takeoff through climb to 39,000ft where the aircraft set up for the first dive point. Sky conditions were clear with winds 330 at 10 knots. The flight then proceeded as follows:

Tim to Impact	Altitude (ft)	Speed (M)	Heading (° mag)	Remarks
	33,000 to 31,000	0.875M (indicated)		Pilot calls "Mark" then conducts single pulses of elevator, aileron and rudder.
	Climb to 39,000ft			Chase reports no mechanical anomalies observed. TM reports all modes well damped. Pilot and TM confer. Next point could be last due to fuel.
T minus 1:17	33,000	0.884M (indicated)	074	All 3 surfaces pulsed. Pilot makes comment "holding full aileron trim and I can't let go". TM clears airplane to 0.894 if flight conditions permitted.

Tim to Impact	Altitude (ft)	Speed (M)	Heading (° mag)	Remarks
T minus 0:50	~30,000 to ground	Increased to > 1.0M	Cycling between 065 and 095 deg	Pilot does not acknowledge speed clearance, instead reporting the aircraft is rolling right and he can't stop it. Chased called "get out, get out." Test aircraft pilot replied he couldn't due to too many g's. Test aircraft completes 6 more rolls while aircraft continues to increase dive angle.
T-0				Telemetry data ends

The main crater was cut almost straight down, about 5 feet, into a sandstone formation. The wreckage was scattered across 9 acres, arrayed 360 degrees around the impact crater. Flight control continuity could not be confirmed due to the severity of the impact damage. The airplane's recorders were located but their condition precluded any data recovery. NTSB report data was obtained from Sino-Swearingen telemetry.

Wind tunnel tests conducted after the accident would reveal decreasing lateral stability with increasing Mach and angle of attack. The aircraft was found to be laterally unstable above 0.83M and any rudder or elevator input made to augment lateral trim could further degrade lateral stability. The wind tunnel results would also indicate the ailerons to be in flow-separated regions between 0.86M and 0.88M. Sino-Swearingen proceeded to make several improvements following the accident. Vortex generators were added to the wings to delay the onset of shock-induced separation. Thicker trailing edge ailerons were installed to improve aileron effectiveness at high Mach number and a high-Mach-number roll spoiler system was devised to augment roll control above Mach 0.835. The single speedbrake panel on each wing was relocated farther outboard to minimize the large pitch-down effects caused by tail lift interference, and the speedbrakes became operational at all airspeeds within the design deployment range. And, finally, the new flutter test aircraft was outfitted with a *high speed drag chute (not present on the accident aircraft) before flutter testing resumed.*[26,27]

So just how hard should a test team push in trying to accomplish tests? Think back to the Gulfstream G650 case. Hindsight can make this look easy but it is much less clear when using foresight. In both the SJ30 and G650 stories we see individual points along a trend. Hindsight tells us these trends were not good. Obviously these two test teams were aware of the issues. When viewing their issues against the objectives of just that one test, both teams decided to go ahead. History now tells us they made the wrong decision. The hindsight advantage tells us these teams simply went too far.

The trick is finding a way to do trend analysis with foresight and not just hindsight. Focusing on both the forest and the trees in real time is something no one human can do very well for very long. For this reason each test program should have people who are involved and informed who are not responsible for day to day operations. These people have the freedom of mind to consider the trends and extrapolate ahead when considering planned tests. When you find such people in your test organization, treat them as your technical conscience and pay special attention to their input.

---

[26] National Transportation Safety Board Factual Report IAD03MA049. http://www.ntsb.gov/aviationquery/brief2.aspx?ev_id=20030515X00671&ntsbno=IAD03MA049&akey=1

[27] Gilbert, Gordon. "Deficient Mach Research Caused SJ30-2 Accident." AINOnline, November 1, 2006. http://www.ainonline.com/aviation-news/aviation-international-news/2006-11-01/deficient-mach-research-caused-sj30-2-accident

8 八

Dealing With Schedule Pressure

Flight test is expensive, and "time is money" as they say. Thus almost no organization that needs to accomplish flight test can afford for it to take any longer than necessary. That leads the development organizations to push for schedule compression in all sorts of ways. Eliminating testing altogether, rearranging tests, attempting tests before preparations are completed: all are fair game. Even when things are working well, the pushers interject trying to make things run still faster. When things get in this mode, no question seems too foolish or brazen to be asked. And given economic realities this situation will always be present to some degree. As many a program manager has said, flight test doesn't make any money -- it only spends money.

> *"What is important is seldom urgent and what is urgent is seldom important."*
> -- *Dwight D. Eisenhower*

So the pressure is always going to be there. We cannot change that fact, and trying to change it is a waste of precious energy. That means the only question is how to respond to it. During the preparations for the initial lunar landing, the Apollo program had its own schedule pressures. The president had publicly committed to have a person standing on the moon before the end of the decade. After making this charge, the president was assassinated. Competition with the Russians was already a driving motivation for the space program. Now honoring this President's legacy made the urge to "get it done" all the more intense.

During the development process, Apollo I burned up on the launch pad killing the three astronauts on board. An investigation committee would be formed to determine proper response to the incident. Prior to that committee's deliberations, mission control director Gene Kranz spoke to his staff and gave them the following charge:

"Space flight will never tolerate carelessness, incapacity and neglect. Somewhere, somehow we screwed up. It could have been a design in build or in test, but whatever it was, we should have caught it. We were too gung-ho about the schedule, and we locked out all of the problems we saw each day in our work. Every element of the program was in trouble and so were we. The simulators were not working, Mission Control was behind in virtually every area, and the flight and test procedures changed daily. Nothing we did had any shelf life. Not one of us stood up and said, `Damnit. Stop.'

I don't know what the Thompson Committee will find as the cause, but I know what I find. We are the cause. We were not ready. We did not do our job. We were rolling the dice, hoping that things would come together by launch day when, in our hearts, we knew it would take a miracle. We were pushing the schedule and betting that the Cape would slip before we did.

From this day forward, Flight Control will be known by two words, `tough' and `competent.' Tough means we will forever be accountable for what we do or what we fail to do. We will never again compromise our responsibilities. Every time we walk into Mission Control, we will know what we stand for. Competent means we will never take anything for granted. We will never be found short in our knowledge and in our skills. Mission Control will be perfect. When you leave this meeting today you will go to your office and the first thing you will do there is to write 'Tough and Competent' on your blackboards. It will never be erased. Each day when you enter the room these words will remind you of the price paid by Grissom, White, and Chaffee. These words are the price of admission to the ranks of Mission Control." [28][29]

So Kranz acknowledged the schedule pressure was not going away. Not wanting to be responsible for causing a slip to the schedule appears to have been a concern for mission control just like it has been for probably

---

[28] Kranz, Gene. <u>Failure is Not an Option: Mission Control from Mercury to Apollo 13 and Beyond</u>. New York: Simon & Schuster, 2000.

[29] Gene Kranz. <u>http://en.wikipedia.org/wiki/Gene_Kranz</u>

every test team in history. How test teams respond to that pressure is critical.

Prior to the accident, technical planning was apparently not allowed to conclude (witness the "shelf life" comment). The risk-mitigating simulators were not working, so their reason-for-being could not be realized. So mission control knew that things were not going right but they chose to proceed anyway. And in so doing mission control passed by their first potential mode of response: the forced halt to testing. They could have just said "no" to specific tests at specific times. Obviously that would have resulted in unwanted attention but when the rest of the wider organization is all primed and ready to go yet a prerequisite is not being met, then the flight test team must have the guts to say "no".

Read the Kranz statement again. Placing a halt on the program is not the only thing he stressed. He also pointed to the need to get ahead of the schedule and accomplish all the necessary preparations so the schedule could be supported. So the more frequent mode of response to schedule pressure is planning ahead and getting preparations made. Yet when a test team comes upon an individual test whose prerequisites are not satisfied they still need to be able to say "no."

> *"Think ahead. Don't let day-to-day operations drive out planning."*
> -- Donald Rumsfeld

Of course accomplishing this is not so simple. Schedules are typically not static. Everyone makes preparations for earlier tests before later tests. Then, when schedules change, the lesser-prepared tests are brought in front of the more-prepared tests and the team is forced to scramble anew to complete preparations for the altered sequence. This is where the scheduling pressure must be engaged -- not resisted or ignored but engaged. There must be some time horizon within which sequence changes are not made. Otherwise test preparations will never have been allowed to complete. The amount of time necessary for that horizon will change based on the particular organizations involved as well as special logistical considerations with the particular test article.

Engaging the schedule pressure does not just involve the chain leading to execution of individual tests. There should also be a functioning technical "conscience" that can see forest in spite of the trees, who can connect the dots and see patterns developing which require additional response. Think back to the story of the Swearingen SJ30 and Gulfstream G650 accidents. In both cases there were technical activities in work which were evaluating the data being seen by testing. And in both cases that evaluation process was not being completed in a time span permitting changed action by the test crews. Consider the following statement by NTSB chairwoman Deborah Hersman in her opening comments to the G650 hearing:

> *"As we delved into the circumstances of the crash, several issues emerged. First, flight testing should not be rushed or compromised. In other accident investigations, we have seen the tragic consequences of schedule pressures or 'get-there-itis.' ... In this investigation we saw an aggressive test flight schedule and pressure to get the aircraft certified. Assumptions and errors were made, but they were neither reviewed nor re-evaluated when new data was collected. Deadlines are essential motivators, but safety must always trump schedule.*
>
> *Second, as aircraft become more sophisticated and more complex, such as the G650, Gulfstream's first fly-by-wire business jet, the planning and flight test environment should respect the aircraft's complexity. That means an abundance of precaution and well-defined procedures. Two prior close calls should have prompted a yellow flag, but instead of slowing down to analyze what had happened, the program continued full speed ahead. Without a Safety Management System or a safety officer, those early warnings were not heeded.*

Third, in all areas of aircraft manufacturing, and particularly in flight testing where the risks are greater, leadership must require processes that are complete, clear and include well-defined criteria. This crash was as much an absence of leadership as it was of lift."[30]

Gulfstream is certainly not alone. One newspaper quoted an unnamed test pilot as saying:

"It doesn't matter who you are — Gulfstream, Bombardier, Airbus. When you are developing a new product, the pressure to get it certified in a timely manner is huge."[31]

Keeping the short term and intermediate-to-long term in full view simultaneously is difficult for any one human to accomplish. Obviously the crews executing individual tests must ensure that every last piece of the short-term sequence is accomplished. For that reason, the longer-term technical view is best taken by individuals not charged with accomplishing individual tests.

Schedule pressure is never going to go away. Keep the focus on doing the right thing in executing tests. Do not allow fighting the schedule pressure to get in the way of accomplishing the tasks at hand. Find a way to establish a time horizon within which schedule changes are not permitted, then plan ahead to make sure things are sufficiently complete when they come within that time horizon. And -- when all else fails -- it is your duty to yourself and to the wider organization to stand up and say "no" if you are being asked to perform a test that simply is not ready to go. No test program has ever benefited from the loss of life or loss of a test article and no test program ever will.

---

[30] Hersman, Deborah A.P. Opening Statement, Aircraft Accident Report - Crash During Experimental Test Flight Gulfstream G650 Roswell NM April 2nd 2011. Washington DC October 10th, 2012.
http://www.ntsb.gov/news/speeches/hersman/daph121010o.html

[31] Mayle, Mary Carr. "NTSB: Warning signs unheeded in Gulfstream G650 crash." Savannah Morning News, October 10th 2012.

Index

72382497R00034

Made in the USA
Columbia, SC
20 June 2017